The Incredible World of Insects

Incredible Bees

By Susan Ashley

Gareth Stevens Publishing

Please visit our website, www.garethstevens.com. For a free color catalog of all our high-quality books, call toll free 1-800-542-2595 or fax 1-877-542-2596.

Library of Congress Cataloging-in-Publication Data

Ashley, Susan.
Incredible bees / Susan Ashley.
p. cm. — (The incredible world of insects)
Includes index.
ISBN 978-1-4339-4577-9 (pbk.)
ISBN 978-1-4339-4578-6 (6-pack)
ISBN 978-1-4339-4575-5 (library binding)
1. Bees—Juvenile literature. 2. Honeybee—Juvenile literature. I. Title.
QL565.2.A743 2012
595.79'9—dc22

2010035220

New edition published 2012 by
Gareth Stevens Publishing
111 East 14th Street, Suite 349
New York, NY 10003

Original edition published 2004 by Weekly Reader® Books
An imprint of Gareth Stevens Publishing

Designer: Daniel Hosek
Editors: Mary Ann Hoffman and Kristen Rajczak

Photo credits: Cover, pp. 1, 9, 15, 17, 19 Shutterstock.com; p. 5 Steve Hopkin/Taxi/Getty Images; pp. 7, 11 © Robert & Linda Mitchell; p. 13 © E. S. Ross/Visuals Unlimited; p. 21 © Bill Beatty/Visuals Unlimited.

Printed in the United States of America

CPSIA compliance information: Batch #CS11GS: For further information contact Gareth Stevens, New York, New York at 1-800-542-2595.

Contents

Boldface words appear in the glossary.

Hives

Large groups of bees live together in homes called hives. Each group has a queen bee. It also has male bees and female bees.

queen

Each bee in a hive has a job. The male bees are called drones. They **mate** with the queen. The queen lays eggs. The female bees are worker bees. They raise the young, gather food, and guard the hive.

Inside the Hive

Worker bees build tiny wax rooms inside the hive. The rooms are called cells. The bees store food in some cells. The queen lays eggs in other cells.

cells

A Bee Comes to Life

Bees begin life as eggs. The eggs **hatch**. A **larva** comes out of each egg. A larva looks like a small worm. A larva does not have legs or eyes.

eggs

A larva changes shape after a few days. It becomes a **pupa**. During this stage, the body of the bee takes shape.

pupa
larva

Body Parts

Bees have six legs. They have three body parts—a head, a **thorax**, and an **abdomen**. A bee's body is covered with tiny hairs.

abdomen
thorax
head

Bees have two pairs of wings. The wings move very quickly. They make a buzzing sound when a bee flies.

Finding Food

Bees fly from flower to flower to find food. They use their tongues like straws to get the flower's **nectar**. They make it into honey.

Bees also collect **pollen** from the flowers. The pollen sticks to the hairs on their bodies. They take it back to the hive to feed to their young.

Glossary

abdomen: the back part of an insect's body

hatch: to come out of an egg

larva: the stage between egg and pupa

mate: to come together to make babies

nectar: sweet liquid from a plant

pollen: powdery grains on a flower

pupa: the stage between larva and adult

thorax: the middle part of an insect's body

For More Information

Books

Hudak, Heather, C. *Bees*. New York, NY: Weigl Publishers, 2009.

Slade, Suzanne. *Bees*. New York, NY: PowerKids Press, 2008.

Web Sites

Beekeeping Links for Kids

www.ncbeekeepers.org/kidlinks.htm

Read facts and play games as you learn about bees.

Bees

42explore.com/bees.htm

Learn facts about bees and find other interesting and fun sites.

Publisher's note to educators and parents: Our editors have carefully reviewed these websites to ensure that they are suitable for students. Many websites change frequently, however, and we cannot guarantee that a site's future contents will continue to meet our high standards of quality and educational value. Be advised that students should be closely supervised whenever they access the Internet.

Index